SL

Standard of Ministry of Water Resources of the People's Republic of China

SL 212—2012

Replace SL 212—98

Design Specification for Hydraulic Prestressed Anchorage

Drafted by:
China Water Northeastern Investigation, Design & Research Co., Ltd.

Translated by:
China Water Northeastern Investigation, Design & Research Co., Ltd.

China Water & Power Press
Beijing 2019

中国水利水电出版社
www.waterpub.com.cn
· 北京 ·

图书在版编目（CIP）数据

水工预应力锚固设计规范 = SL212-2012 Design Specification for Hydraulic Prestressed Anchorage : 英文 / 中华人民共和国水利部发布. -- 北京 : 中国水利水电出版社, 2019.9
ISBN 978-7-5170-8044-2

Ⅰ. ①水… Ⅱ. ①中… Ⅲ. ①预应力加筋锚固－设计规范－英文 Ⅳ. ①TU757.2-65

中国版本图书馆CIP数据核字(2019)第210680号

书　　名	**Design Specification for Hydraulic Prestressed Anchorage SL 212—2012**
作　　者	中华人民共和国水利部　发布
出版发行	中国水利水电出版社 （北京市海淀区玉渊潭南路 1 号 D 座　100038） 网址：www.waterpub.com.cn E-mail：sales@waterpub.com.cn 电话：(010) 68367658（营销中心）
经　　售	北京科水图书销售中心（零售） 电话：(010) 88383994、63202643、68545874 全国各地新华书店和相关出版物销售网点
排　　版	中国水利水电出版社微机排版中心
印　　刷	清淞永业（天津）印刷有限公司
规　　格	140mm×203mm　32 开本　1.5 印张　40 千字
版　　次	2019 年 9 月第 1 版　2019 年 9 月第 1 次印刷
定　　价	**180.00** 元

Introduction to English Version

Department of International Cooperation, Science and Technology (DICST) of Ministry of Water Resources, P. R. China has the mandate of managing the formulation and revision of water technology standards in China.

Translation of this standard from Chinese into English is organized by DICST in accordance with due procedures and regulations applicable in China.

The English version of this standard is identical to its Chinese original SL 212—2012, which was formulated and revised under the auspices of DICST.

Translation of this standard is undertaken by China Water Northeastern Investigation, Design & Research Co. , Ltd.

Translation task force includes SU Jialin, CHEN Liqiu, WANG Deku, MA Donghe, GAO Yin, LI Yanping, ZHANG Yu, TAN Chun, LU Puguang, XIAO Yang, ZHU Haibo and LIU Zhixin.

This standard is reviewed by QIAO Shishan, WU Nongdi, WANG Yicheng , XU Zeping, HUANG Jinchi, ZHENG Cuiying, BAO Shujun, CHEN Shaosong, LEI Xingshun, WANG Yujie, FAN Shaopeng, YANG Haiyan, WANG Baocheng and CHEN Jiansu.

Department of International Cooperation, Science and Technology
Ministry of Water Resources, P. R. China

Foreword

The revision of SL 212—98 *Design Specification for Hydraulic Prestressed Anchorage* was in accordance with the *Plan on Formulating and Revising the Sectoral Standard of the Ministry of Water Resources* and SL 1—2002 *Specification for the Drafting of Technical Standards of Water Resources* issued by the Ministry of Water Resources.

This standard consists of 8 chapters, 19 sections, 153 articles, and 1 annex. The major contents of this standard include general provisions, terms and symbols, general requirements, design of anchorage system, anchorage design of slopes, anchorage design of tunnels and caverns, anchorage design of hydraulic structures, design and test of safety monitoring, and test of tendon bearing capacity.

This edition of standard revises and supplements the following contents: adding the contents of multi-unit compression grouted tendon, tension-compression-combined tendon, and anchorage design of soil slopes, rock bolt crane beam, and concrete lining of hydraulic tunnel by ring prestressed tendon; revising " reparation and reinforcement of hydraulic structures" to "anchorage of concrete dams" and revising Annex A "The Provisions of Prestressed anchorage test" to "Test of tendon bearing capacity"; and removing Annex B of the previous edition.

Compared with the previous one, this edition of standard reserves 12 articles, revises 71 articles, and adds 70 articles, embodying the innovative progress in prestressed anchorage technology.

This standard replaces SL 212—98.

This standard is approved by Ministry of Water Resources of the People's Republic of China.

This standard is interpreted by General Institute of Water Resources and Hydropower Planning and Design, Ministry of Water Resources.

This standard is drafted by China Water Northeastern Investigation, Design & Research Co., Ltd.

This standard is published and distributed by China Water & Power Press.

Chief drafters of this standard are ZHAO Changhai, SU Jialin, SU Ping, GAO Yin, DU Guowen, CHEN Lei, CHEN Liqiu, JING Jianwei and ZHENG Yifang.

The leading expert of the technical review meeting of this standard is WEN Xuyu.

The format examiner of this standard is CHEN Dengyi.

Contents

1 General Provisions

1.0.1 This standard is formulated with aims of keeping pace with the progress of prestressed anchorage technology and standardizing its design requirements so as to make the prestressed anchorage design safe and applicable, economically viable, and technologically advanced.

1.0.2 This standard is applicable to the design of prestressed anchorage to be used for foundations, slopes, tunnels and caverns, and hydraulic concrete structures of water and hydropower projects.

1.0.3 New technologies, new equipment and new materials shall be introduced in design of the prestressed anchorage.

1.0.4 The standards quoted in this standard mainly include:

GB 175 *Common Portland Cement*

GB/T 700 *Carbon Structural Steels*

GB/T 5223 *Steel Wire for Prestressing of Concrete*

GB/T 5224 *Steel Strand for Prestressed Concrete*

GB/T 14370 *Anchorage, Grip and Coupler for Prestressing Tendons*

GB/T 20065 *Screw-thread Steel Bars for the Prestressing of Concrete*

GB 50021 *Code for Investigation of Geotechnical Engineering*

GB 50487 *Code for Engineering Geological Investigation of Water Resources and Hydropower*

SL 105 *Specifications for Anticorrosion of Hydraulic Steel Structures*

SL 191 *Design Code for Hydraulic Concrete Structures*

SL 211 *Code for Design of Hydraulic Structures against Ice and Freezing Action*

SL 253 *Design Code for Spillway*

SL 265 *Design Specification for Sluice*

SL 279 *Specification for Design of Hydraulic Tunnel*

SL 379 *Design Specification for Hydraulic Retaining Wall*

SL 386—2007 *Design Code for Engineered Slopes in Water Resources and Hrdropower Projects*

1.0.5 For hydraulic prestressed anchorage design, provisions stipulated not only in this standard but also in the prevailing relevant standards of China shall be complied with.

2 Terms and Symbols

2.1 Terms

2.1.1 Prestressed anchorage

A technical measure of installing the prestressed tendons in rock mass or concrete structures and applying stressing force to them so as to stabilize the rock mass or concrete structures, or improve their stress status.

2.1.2 Tendon

A member made up of a single or a plurality of steel wires or strands in a certain pattern.

2.1.3 Anchor

A rod made of metal or resin material for anchorage works.

2.1.4 Prestressed tendon

Tendon to which the prestressed force has been applied. Both prestressed tendon and tensioned anchor are referred to as prestressed tendon in this specification.

2.1.5 Permanent prestressed tendon

Prestressed tendon installed in permanent works for permanent use.

2.1.6 Temporary prestressed tendon

Prestressed tendon installed in permanent works or temporary works and for not more than 2 years service life.

2.1.7 Tensioned anchor

The anchor to which the prestressed force has been applied.

2.1.8 Tendon element

A prestressed tendon set, including fixed length, free length, anch or head and other parts.

2.1.9 Fixed length

The internal force holding section that bears the stressing force of the tendon by bonding the tendon onto the object to be anchored through grouted materials or mechanical devices.

2.1.10 Free length

The section which can be free to be elongated during stressing.

2.1.11 Anchor head

The supporting and clamping device designed to impose the stressing force and lock the tendon.

2.1.12 Bonded prestressed tendon

A prestressed tendon in which there is no relative movement between free length and the object to be anchored after completion of locking and grouting.

2.1.13 Unbonded prestressed tendon

A prestressed tendon in which there is relative movement between free length and the object to be anchored even though locking and grouting has been done to lock off the tendon.

2.1.14 Utilization factor of the strength of tendon materials

The ratio of average stress of tendon material to the characteristic value of the tendon tensile strength when the tensile stress inside the prestressed tendon reaches its design value.

2.1.15 Design stressing force

The design stressing force applied to each tendon determined according to the anchorage design requirement, considering the safety margin, the prestress loss due to rock rheology, creep of concrete and relaxation of steel wire.

2.1.16 Design anchorage force

Permanent force to be retained in the tendon after deducting the prestress loss due to various factors.

2.1.17 Extra stressing force

The stressing force which is increased in construction period to exceed the design anchorage force so as to compensate the prestress loss due to the relative movement between tendon and borehole, the compression of anchorage device and the draw-in of tendon.

2.1.18 Initial tensioning

Applying of initial force before tensioning to tighten the steel wires/strands of all bundles in order to bear force evenly.

2.1.19 Compensatory tensioning

The re-applying of stressing force to the tendon to compensate the prestress loss after locking off.

2.1.20 Tension grouted tendon

A prestressed tendon whose grout sleeve of fixed length will be in a state of tension when it is in service.

2.1.21 Compression grouted tendon

A prestressed tendon whose grout sleeve of fixed length will be in a state of compression when it is in service.

2.1.22 Multi-unit tension grouted tendon

A tension grouted tendon with its fixed length composed of several different lengths of steel wires/strands bonded to different positions of the borehole to disperse the tension.

2.1.23 Multi-unit compression grouted tendon

A compression grouted tendon with its fixed length composed of several bearing elements to disperse the pressure.

2.1.24 Tension-compression-combined tendon

A prestressed tendon composed of several groups of compression and tension grouted tendon.

2.1.25 Draw-in length

The length of draw-in of steel wires/strands in the anchor's free length when locking the prestressed tendon.

2.2 Symbols

L—length of prestressed tendon;

L_1—length of fixed length;

L_2—length of free length;

L_3—length of steel strand on the anchor head;

P_m—extra stressing force of a single prestressed tendon;

C—bond strength between the grouted material and borehole;

D—diameter of borehole;

n—number of steel wires/strands of a single tendon;

A—control area of a single tendon;

q_1—design capacity of a single tendon;

P_1—supporting force per unit area applied by the prestressed tendon;

P_2—supporting force per unit area applied by tensioned anchors;

P_3—supporting force per unit area applied by steel mesh shotcrete;

P_4—resisting force per unit area of surrounding rock;

P_i—minimum supporting force per unit area necessary to maintain the stability of surrounding rocks;

β—optimal anchorage angle (the angle included between the tendon axis and the horizontal plane);

α—dip of sliding plane (weak discontinuity);

φ—internal friction angle;

σ_x—tensile stress on calculated cross-section;

σ_k—control values of tensile stress of the steel strand during tensioning;

σ_1—retained tensile stress after deducting the deflector's friction loss;

a—coefficient of deflector's friction loss;

σ—effective tensile stress along the streel strand;

k—friction coefficient (wobble coefficient) of a steel strand;

μ—friction coefficient between the steel strand and the sheath;

θ—angle included between the tangent of tensioning end and the tangent of the curved duct at the calculated cross-section;

f_{ptk}—characteristic values of steel strand's tensile strength.

3 General Requirements

3.1 Basic Data

3.1.1 The following basic data shall be available for prestressed anchorage design:

1 Structure class and project layout.

2 Basic parameters, load combinations and operation features of hydraulic structures.

3 Topographic and geological conditions of the anchorage area.

4 Construction conditions.

5 Physical and mechanical properties of related materials.

3.1.2 The geological investigation of the area to be reinforced with prestressed tendons shall be carried out in accordance with GB 50487. And the following data shall be available for the anchorage design:

1 Geological map and cross-sections of anchorage works.

2 Extent and boundary conditions of rock/soil in anchorage area.

3 Data on quality of rocks, attitude of the main tectonic structure, combination relationship among various discontinuities, geostress, and groundwater.

4 Physical and mechanical properties and parameters of rock/soil in the areas to be anchored, and values of c and ϕ of discontinuities along which instability is prone to take place.

5 Aggressive ion indexes of the groundwater of the anchorage area, etc.

3.1.3 For an important location to be anchored, necessary physical and mechanical parameters shall be obtained by in-situ tests.

3.2 Materials

3.2.1 High tensile and low relaxation steel wires, steel strands, steel bars, or hollow prestressed anchor made of high tensile materials may be selected as tendon materials according to project nature scale, and anchoring conditions.

3.2.2 The mechanical properties of high tensile steel strands/wires used as tendon material shall conform to GB/T 5224 and GB/T 5223 respectively.

3.2.3 For unbonded prestressed bars used as tendons, the material performance shall meet the following requirements:

1 The material property of steel used for making unbonded prestressed bars shall comply with GB/T 5224 or GB/T 5223.

2 Special anti-corrosion grease shall be applied on the unbonded prestressed bars, and shall be:

1) Neither flow, nor crack, nor embrittle under the temperature ranging from −20 ℃ to +70 ℃.

2) With stable chemical performance during its service life.

3) Non-aggressive to the adjacent materials.

4) Watertight.

5) Of high corrosion-resistance.

6) With good lubrication and low friction.

3 Polyethylene or polypropylene shall be used as the sheath of the unbonded prestressed bars, but polyvinyl chloride (PVC) shall be prohibited. The sheath shall meet the following requirements:

1) Be serviceable under the temperature ranging from −20 ℃ to +70 ℃, not embrittle at low temperature, and have stable chemical performance at high temperature.

2) Have adequate ductility and toughness.

3) Be non-aggressive to adjacent materials.

4) Be good in waterproofing.

3.2.4 The properties of high-tensile deformed bars used as prestressed tendon material shall comply with GB/T 20065.

3.2.5 The performance and quality of prestressed anchor head and prestressed reinforcement coupler shall comply with GB/T 14370.

3.2.6 Portland cement or ordinary portland cement shall be used for grouting of the fixed length and filling of prestressed tendon hole, and its quality shall meet the requirements of GB 175. When ground water is aggressive, special cement shall be used.

3.2.7 Materials of concrete bearing block shall meet requirements of SL 191 and those of steel bearing plate shall comply with GB 700.

3.3 Basic Requirements for Anchorage Design

3.3.1 The anchorage design shall include:

1 Determining the anchoring extent and depth.

2 Selecting the type of anchorage.

3 Calculating the anchorage force.

4 Determining the number of prestressed tendons and selecting the scheme of layout.

5 Determining the configuration and parameters of the tendons.

6 Drafting layout of anchorage works and configuration of prestressed tendons, and preparing technical specifications.

7 Proposing additional treatment measures for the anchorage area with complicated geological conditions.

8 Performing monitoring design for anchorage.

3.3.2 The anchorage extent and forces of prestressed tendons shall be determined based on the stability analysis according to the engineering geological data, the position, attitude and mechanical properties of discontinuities, or the stress state of the structure, etc.

3.3.3 The design stressing force and extra stressing force of a prestressed tendon shall be determined according to the following factors:

1 The total anchorage force necessary to ensure the safe operation of the anchored structure.

2 The possible prestress loss due to rock rheology, concrete creep, and relaxation of steel.

3 Mechanical parameters of anchored object and grouted materials.

4 Mechanical parameters of prestressed tendon materials.

5 Type of gripping devices, capacity of stressing device and conditions of construction site.

3.3.4 The number of prestressed tendons shall be determined based on the total anchorage force and the design anchorage force of a single prestressed tendon.

3.3.5 For prestressed tendons used for reinforcement of slopes, tunnels and caverns, and foundations, the length shall be determined according to the position of potential failure plane and the fixed length in the stable rock mass, etc.

3.3.6 For prestressed tendons used for reinforcement of hydraulic structures, the length shall be determined according to the structure size and stress analysis.

3.3.7 The prestressed tendons used in rock mass shall be arranged according to the following principles:

1 The arrangement of tendons shall be conducive to producing a uniform anchoring force. They shall be patterned as square, rectangle, diamond or staggered, based on the number of tendons, construction conditions and technical requirements.

2 The longitudinal direction of the prestressed tendon should be inclined according to the optimal anchoring angle.

3 For group anchoring, the fixed length of adjacent prestressed tendons should be staggered, and the inclinations of tendons may be adjusted if necessary.

3.3.8 The prestressed tendons used in hydraulic structures shall be arranged according to the following principles:

1 The arrangement of prestressed tendons in gate pier shall be determined by comprehensive comparisons according to the structural shape of the pier, the shape of bearing block, stress analysis of the pier, and construction conditions.

2 The arrangement of annular prestressed tendons used for concrete lining shall be determined based on the stress analysis, type of tendon material, and construction conditions.

3 The arrangement of prestressed tendons used in a concrete dam and its foundation, sluice/gate chamber, stilling basin/pond, retaining wall, and other hydraulic structures shall be determined in accordance with stability analysis and stress analysis.

3.3.9 For permanent prestressed tendons, the protection against corrosion and rusting shall be designed based on the importance of the project, the surrounding medium, the chemical nature of the seepage water, etc.

3.3.10 For an important project or an important part of a project, a certain number of testing tendons shall be provided to determine the stressing force, the extra stressing force, the possible prestress losses, and the possible compression deformation of the anchored object, etc., so as to justify the design parameters.

4 Design of Anchorage System

4.1 Selection of Tendon Element

4.1.1 The type of tendon element shall be selected through comprehensive comparisons based on the service life, the design anchorage force of a single tendon, the arrangement pattern of tendons, and the construction conditions.

4.1.2 Fixed length, free length, anchor head and various couplings of the prestressed tendon shall be designed according to the principle of equal strength.

4.1.3 The length of prestressed tendon depends on the condition of the anchored rock/soil with the length embedded in the stable rock/soil mass no less than the designed length of fixed length. The total length of the prestressed tendon is the sum of fixed length, free length, anchor head and exposed length.

4.1.4 For tension grouted tendon, compression grouted tendon, and tension-compression-combined tendon, every steel wire/strand shall be of the same length. For both multi-unit tension grouted tendon and multi-unit compression grouted tendon, the length of every steel wire/strand shall be determined according to the number of units.

4.1.5 For permanent anchorage works, the prestressed tendons with bond anchorage shall be selected. Tendons for mechanical anchorage may be adopted under the following conditions: the design anchorage force of a single prestressed tendon is less than 1000 kN and the compressive strength of rock in the anchorage zone is greater than 60 MPa, prompt anchorage is required or using of bond anchorage is infeasible.

4.1.6 When the designed length of fixed length exceeds 10m, such anchorage type as multi-unit tension grouted tendon, multi-unit compression grouted tendon or tension-compression-combined tendon shall be adopted.

4.1.7 The prestressed tendons shall be selected as per the following principles:

1 For permanent prestressed tendons, high tensile and low relaxation steel strands/wires should be adopted.

2 When conditions permit, bonded tendons shall be preferred. For prestressed tendons used for monitoring and requiring for future compensatory stressing, unbonded steel strands shall be adopted.

3 For prestressed tendons requiring a certain degree of stiffness or special installation, high tensile deformed steel bars may be adopted.

4 When the rock mass in the anchorage zone is too fractured to form holes, self-drilling tensioned anchor may be adopted.

4.2 Design of Tendon Element

4.2.1 The design of tendon element shall include: selecting materials of tendon element, determining the design anchorage force, the design stressing force, the number of tendons, the configuration of tendons (the location of spacer and centralizer, the arrangement of vent tube, grout tube, ring packer, etc.), the diameter of borehole and the protective measures of the tendon element, etc.

4.2.2 The utilization coefficient of the steel strength should not exceed 60% when using prestressed steel wires/strands as tendon element, nor 65% when using high tensile deformed steel bars or high tensile self-drilling bars.

4.2.3 For reinforced rock/soil potential to develop further significant deformation or displacement according to the geological data, the utilization coefficient of the prestressed steel strength shall be reduced.

4.2.4 The spacers shall be set along the length of tendon. The spacing of spacers should not exceed 4.0 m for tendon with a high inclination nor 2.0 m for tendon with a low inclination.

4.2.5 The diameter of the spacer shall be 20 mm less than that of the borehole. The number of holes in spacers for the passage of wires/strands shall be the same as that of the steel strands, and steel strands in the holes shall be kept straight and smooth. One or more additional holes for the passage of grout and vent tubes shall be provided in the spacer.

4.2.6 The spacer may be made of steel plate or alternative materials. Its texture shall comply with relevant specifications and its thickness should not be less than 10 mm.

4.2.7 After filling grout, the thickness of grout or mortar cover for protecting tendons shall be greater than 20 mm.

4.3 Design of Fixed Length

4.3.1 The anchoring capacity of bond fixed length shall be greater than the extra stressing force of the prestressed tendons. The length of fixed length may be calculated with Formula (4.3.1-1) and verified with Formula (4.3.1-2). The safety factor of fixed length may be selected from Table 4.3.1-1. For an important project or anchoring works with complicated rock conditions, it shall also be justified by field pullout tests.

$$L_1 = k\frac{P_m}{\pi DC} \tag{4.3.1-1}$$

$$L_1 = k\frac{P_m}{\pi d c_1 n} \tag{4.3.1-2}$$

Where:

L_1 = length of fixed length, mm;

P_m = extra stressing force of a single prestressed tendon, N;

k = safety factor of the length of fixed length, see Table 4.3.1-1;

D = diameter of borehole, mm;

C = bond strength between the grouted material and borehole, and it may be selected from Table 4.3.1-2 when no test data is available;

c_1 = bond strength between the grouted material and steel wires/strands, and it may be taken as 2.0 MPa;

d = diameter of a single bundle of steel wires/strands;

n = number of steel wires/strands of a single tendon.

Table 4.3.1-1 Safety factors of bond fixed length

Anchorage nature and hole direction	Permanent tendon		Temporary tendon	
	Up hole	Down hole	Up hole	Down hole
Safety factor k	2.0	1.5	1.6	1.2

Table 4.3.1-2 Bond strength between grout/mortar and rock

Classification of rock	I	II	III	IV	V
Bond strength (MPa)	1.5	1.5-1.2	1.2-0.8	0.8-0.3	≤0.3

4.3.2 The fixed length should not be longer than 10m, otherwise one or more of the following measures should be taken: improving the quality of rock mass, expanding the borehole diameter of the fixed length, using multi-unit tension grouted tendons or multi-unit compression grouted tendons, etc., and a special design for the fixed length shall be made.

4.3.3 Cement slurry, cement mortar or resin shall be selected as grouted material according to factors such as the requirements of the anchorage works, the strength of rock mass in the zone of fixed length, the structure of the fixed length. The compressive strength of cement slurry/mortar should not be lower than 35 MPa, and that of resin should not be lower than 50 MPa.

4.3.4 For either multi-unit tension grouted tendons or multi-unit compression grouted tendons, the number of fixed units and the lengths of corresponding steel strands shall be determined according to such factors as the total length of the tendon, the geological conditions at the fixed length zone, the diameter of the borehole, and the compressive strength of the grout column. The anchorage force of each fixed unit shall be calculated respectively. The total anchorage force of all fixed units shall be greater than the design anchorage force of a single tendon.

4.3.5 For compression grouted tendons or multi-unit compression grouted tendons, the bearing plate and other bearing mechanism at the fixed units shall be mature products made by professional manufacturers, and shall meet quality requirements.

4.4 Design of Anchor Head

4.4.1 Anchor head shall include a concrete bearing block, a bearing plate, a stressing head and the sealing protection. For a tendon with monitoring requirement, an instrument for measuring the anchorage force shall be provided.

4.4.2 The length of the steel strand at the anchor head shall be the sum of a margin of 0.3 m to 0.5 m plus the total height of concrete bearing block, bearing plate, mounted stressing head, detachable stressing head and jack. For a tendon with monitoring requirement, the height of the pressure transducer shall also be included.

4.4.3 The bearing capacity of each component shall be matched with the extra stressing force of a tendon, and its material property shall comply with GB/T 14370.

4.4.4 The draw-in length of steel wires/strands shall not exceed 5 mm after locking off the tendons.

4.4.5 The type and size of concrete bearing block shall be designed according to the capacity of a tendon and geological conditions. The bearing surface of concrete bearing block shall be perpendicular to the stressing direction of prestressed tendon. For concrete bearing block, the strength grade should not be lower than C30 and the frost resistance shall meet relevant requirements of SL 211. Two additional holes shall be provided in concrete bearing block to allow the passage of grout and vent tubes.

4.4.6 Steel bearing plate shall be placed tightly on the bearing surface of the concrete bearing block. The contact surface of the steel bearing plate with the mounted stressing head shall be kept smooth and clean. The thickness of steel bearing plate may be determined according to the capacity of a tendon, but should not be less than 20 mm.

4.5 Protection Design of Prestressed Tendon

4.5.1 Special protection design against chemical corrosion, stress corrosion and electrostatic corrosion shall be conducted according to the importance of project, the groundwater properties of anchorage zone and design anchorage force.

4.5.2 Aggressivity of circumstances to prestressed tendons may be divided into four degrees: non-aggressive, mildly aggressive, moderately aggressive and very aggressive. The degrees of aggressivity may be assessed by referring to GB 50021, and protection design shall be conducted according to Table 4.5.2.

4.5.3 Where groundwater is affluent and aggressive, consolidation grouting may be performed before installation of prestressed tendons.

Table 4.5.2 Design criteria for prestressed tendon protection

Aggressivity	Temporary prestressed tendon	Permanent prestressed tendon
Non-aggressive or mildly aggressive	A	C
Moderately aggressive	B	C
Very aggressive	C	D

Notes: A—Liquid protection, limewash, and cosmoline.
B—Plastic protection, hydrogel, resin, anti-rusting grease, etc.
C—Rigid protection, cement grout and cement mortar, corrugated sheath for unbonded prestressed tendons.
D—Double protection, consolidation grouting of the entire hole, corrugated sheath grouted with special mortar/grout.

4.5.4 For the presence of aggressive chemicals in the stratum of anchorage area, unbonded prestressed steel strands are preferred and shall be enclosed by corrugated sheath filled with mortar/grout. The thickness of the mortar/grout outside the corrugated sheath should not be less than 10 mm. The mortar/grout shall be of chemical resistance.

4.5.5 Materials and agents used for anti-corrosion and anti-rusting treatment of the tendon element shall be free of nitrates, sulfite, or thiocyanate. The content of chlorine ion in cement shall not exceed 0.02% of the cement in weight.

4.5.6 The cement of mortar/grout used for filling and as grouting materials shall comply with GB 175.

4.5.7 The corrosion protection of the unbonded prestressed tendon shall comply with Article 3.2.3.

4.5.8 Insulation or cathode protection shall be performed in the presence of stray current in the prestressed anchorage area.

4.5.9 Anti-rusting treatment of tendon element shall be conducted before installation of prestressed tendons, and that of the anchor head shall be done after installation of prestressed tendons. The anchor head shall be sealed immediately after locking off of the tendon and completion of grouting. The anti-rusting treatment of prestressed tendons shall comply with SL 105.

4.5.10 Protection for anchor head of unbonded prestressed tendon shall comply with the following provisions:

1 When stressing and locking the permanent prestressed tendon, the anchor head shall be timely coated with anti-corrosion compound and covered with concrete of not less than 100 mm in thickness.

2 The void within the cap for protecting the anchor head shall be filled with anti-corrosion compound.

4.6 Design of Stressing Procedure

4.6.1 Before stressing a prestressed tendon, the steel strands shall be initially tensioned and locked one by one. The stressing force applied for the initial tensioning may be 10% of the design stressing force.

4.6.2 Extra stressing may be performed if necessary. Extra stressing force should be 110%, but should not exceed 115% of the design stressing force.

4.6.3 The stressing procedure shall be in compliance with the following provisions:

1 The stressing force shall be applied to the extra stressing force progressively in cycles. Generally, the application may be completed in four or five cycles.

2 Each stage's stressing force shall be held for 5 min, and the actual elongation of prestressed tendons shall be measured.

3 When the extra stressing force has been reached, it shall be held for 5min and elongation of tendon shall be measured simultaneously. The tendon shall be then locked off. Thereafter, the actual draw-in of tendon shall be measured.

4.6.4 For the following locations to be anchored with prestressed tendons, specific design of stressing procedure shall be made according to in situ tests or monitored results:

1 Large tunnels and caverns.

2 Swelling rock or soil strata.

3 Rock strata with high geostress.

5 Anchorage Design of Slopes

5.1 Anchorage of Rock Slopes

5.1.1 The stability of rock slopes reinforced by prestressed tendons shall comply with Article 3.4.2 of SL 386—2007. For important anchorage works whose slope failure may result in heavy losses, the safety factor shall be greater than 1.5 through justification.

5.1.2 The arrangement of prestressed tendons shall be in accordance with according to the position and attitude of weak discontinuities prone to instability, the extent of reinforcement, and the capacity of a single tendon.

5.1.3 Prestressed tendons shall be spaced uniformly 3 m to 6 m apart.

5.1.4 Anchorage angle of prestressed tendons should be determined as per the following principles:

1 It should be calculated with Formula (5.1.4):

$$\beta=\alpha-\left(45^{\circ}+\frac{\varphi}{2}\right) \tag{5.1.4}$$

Where:

β= optimal anchorage angle (the angle included between the tendon axis and the horizontal plane);

α= dip of sliding plane (weak discontinuity);

φ= internal friction angle.

2 It should be adjusted to $\beta\leqslant-10^{\circ}$ or $\beta\geqslant+10^{\circ}$ when the optimal anchorage angle is $-10^{\circ}<\beta<+10^{\circ}$.

3 The variation of the optimal angle shall be adjusted through technical and economic comparisons when the arrangement as per the optimal anchor angle is difficult to implement.

5.1.5 When the rock mass under the anchor head fails to accommodate the bearing capacity of the tendon, such measures as increasing the size of concrete bearing block and placing concrete lattice beams may be taken.

5.1.6 For rock slope anchorage zone, the water intercepting and drainage shall be designed as per the relevant provisions of SL 386—2007.

5.2 Anchorage of Soil Slopes

5.2.1 The stability of soil slopes reinforced by prestressed tendons shall comply with Article 3.4.2 of SL 386—2007.

5.2.2 If the sliding plane is on the interface of soil mass and rock mass, the fixed length shall be set into a stable rock.

5.2.3 For sliding plane within soil mass, the fixed length shall be arranged into a stable soil mass, and its length shall be adequate to withstand the extra stressing force, otherwise, such measures as increasing the diameter of cross-section of the fixed length or using multi-unit tension grouted tendons or multi-unit compression grouted tendons shall be taken.

5.2.4 The anchorage angle shall be determined in accordance with Article 5.1.4 of this standard.

5.2.5 For free length in soils, the spacing between spacers should not exceed 2.0 m.

5.2.6 Concrete bearing blocks shall be integrated with supporting structures such as lattice beams and retaining walls.

5.2.7 For soil slope anchorage zone, water intercepting and the drainage shall be designed as per the relevant provisions of SL 386—2007.

6 Anchorage Design of Tunnels and Caverns

6.1 Anchorage of Tunnels and Caverns

6.1.1 Tunnels and caverns with large areas of squeeze-shear fractured zone and plastic deformation or unstable rock wedge/block may be reinforced by prestressed tendons.

6.1.2 The supporting force per unit area applied by the prestressed tendon may be calculated with Formula (6.1.2):

$$P_1 = q_1 / A \qquad (6.1.2)$$

Where:

P_1 = supporting force per unit area applied by the prestressed tendon, MPa;

q_1 = design capacity of a single tendon, N;

A = control area of a single tendon, mm^2.

6.1.3 For tunnels and caverns reinforced with prestressed tendons, tensioned anchors, and steel mesh shotcrete, the stability safety factor of surrounding rocks shall be satisfy Formula (6.1.3):

$$K \geqslant (P_1 + P_2 + P_3 + P_4) / P_i \qquad (6.1.3)$$

Where:

K = stability safety factor of surrounding rocks, taken as 1.5 to 1.8 in accordance with the importance of the project;

P_i = minimum supporting force per unit area necessary to maintain the stability of surrounding rocks, MPa;

P_1 = supporting force per unit area applied by prestressed tendons, MPa;

P_2 = supporting force per unit area applied by tensioned anchors, MPa;

P_3 = supporting force per unit area applied by steel mesh shotcrete, MPa;

P_4 = resisting force per unit area of surrounding rock, MPa.

6.1.4 Prestressed tendons shall go through the squeeze and shear fractured zone or the plastic deformation zone, the fixed length shall be arranged in the elastic zone and its length shall conform to Article 4.3.1 of this standard.

6.1.5 The spacing between prestressed tendons should not exceed half the length of the free length, while the minimum spacing should not be less than 4 m.

6.1.6 The capacity of prestressed tendons at the vault shall be determined as per the total weight of potential collapse block.

6.1.7 The capacity of prestressed tendons for reinforcing sidewall shall be determined based on the requirements for rock slopes, taking account of the attitude combination of joints and fissures and the interlocking of rocks surrounding the potential collapse block.

6.1.8 Wall-penetrating prestressed tendons shall be preferred for anchoring the rock wall between two adjacent large tunnels and caverns.

6.2 Anchorage of Rock Bolt Supported Crane Beam

6.2.1 The rock bolt supported crane beam may be reinforced on the rock wall with prestressed tendons/bolts.

6.2.2 The design capacity of prestressed tendons may be determined by rigid body limit equilibrium method or elastic-plastic finite element method. For rigid body limit equilibrium method, the number of prestressed tendons for one meter long beam may be calculated with Formula (6.2.2):

$$KF = A_s f_y \qquad (6.2.2)$$

$$f_y = 0.8 f_{ptk}$$

Where:

K = Safety factor, taken as 2.0 for Class 1 structures, 1.8 for Class 2 structures, and 1.5 for Class 3 structures;

F = design capacity of prestressed tendon, N;

A_s = cross-sectional area of prestressed tendon, mm^2;

f_{ptk} = characteristic value of steel strand's tensile strength, N/mm^2.

6.2.3 High-tensile deformed bar should be used as prestressed tendons.

6.2.4 The length of the fixed length shall be calculated with Formula (4.3.1-1) plus 1.0 m long of relaxation zone of surrounding rocks.

6.2.5 The angle included between the prestressed tendon and horizontal plane should be 15° to 25°, and it may be set as required under special conditions.

6.2.6 The working performance of prestressed tendons shall be monitored.

7 Anchorage Design of Hydraulic Structures

7.1 Anchorage of Concrete Dams

7.1.1 When prestressed tendon is used to reinforce concrete dam foundation and dam body, analysis of stress and stability against sliding and overturning shall be performed for different objects as per corresponding standards so as to determine the reinforcement extent and the design anchorage force.

7.1.2 According to site and construction conditions, the axis orientation of prestressed tendons used to reinforce dam foundation shall be determined through technical and economic comparisons.

7.1.3 In the presence of discontinuities with gentle dip in the dam foundation, the length and arrangement of prestressed tendons shall be determined according to the location and attitude of discontinuities as well as structure layout and construction conditions. When the design anchorage force of a single prestressed tendon exceeds 3000 kN, multi-unit tension grouted tendons or multi-unit compression grouted tendons may be used.

7.1.4 The prestressed tendons used to reinforce dam foundation shall be protected in terms of Grade C as specified in Article 4.5.2, and Grade D when the groundwater is aggressive.

7.1.5 For the dam foundation with fissures or weak and cracked rocks, consolidation grouting shall be conducted prior to anchorage. The consolidation grouting design shall comply with corresponding standards.

7.1.6 For prestressed tendons used to reinforce the dam body with defects like cracks, the anchorage force selected shall be such as to accommodate the strength of the original structures. The fixed lengths of adjacent tendons shall be staggered.

7.1.7 When grouting is required after the dam cracks have been anchored, the grouting pressure shall be under control.

7.2 Anchorage of Prestressed Concrete Piers

7.2.1 When the total design thrust acting on a radial gate exceeds 25000 kN, the prestressed concrete piers may be used. The structural design of prestressed concrete piers shall comply with SL 191.

7.2.2 The structural mechanics method may be adopted for calculation of prestressed concrete pier and the 3D finite element method for verification. When necessary, the structure model test may be adopted for justification.

7.2.3 The anchorage design of prestressed concrete pier shall include the following:

1 The design anchorage force and the arrangement of main tendons.

2 The design anchorage force and arrangement of auxiliary tendons.

7.2.4 The arrangement of main tendons shall meet the following requirements:

1 The resultant force of prestressed tendons shall pass through the trunnion center, the main tendons in the pier shall be fanned in elevation view along the thrust direction of the radial gate, and the angle included between the tendons and the thrust shall be $-10°$ to $+10°$. The main tendons shall be staggered in length.

2 The main tendons in the middle pier should be arranged symmetrically in plan, while those in the abutment pier and the pier separated with joint should be arranged asymmetrically. The main tendons shall be arranged parallel to the pier's face or the angle between them should not exceed 3°. The main tendons should be near the pier face with a protection distance of not less than 500 mm. The spacing of main tendons should not be less than 400 mm.

7.2.5 A certain number of auxiliary tendons shall be set in the supporting structure of piers.

7.2.6 The diameter of the hole shall be dependent on the diameter of the tendon and the protective cover. The protective cover thickness should not be less than 20 mm for bonded steel strands nor 10 mm for unbonded steel strands. The pre-embedded corrugated duct or steel pipe should be used as the hole.

7.2.7 The concrete strength grade of anchorage zone shall not be lower than C30, nor that of anchor block and support and so on should be lower than C40.

7.3 Anchorage of Gate Chamber, Stilling Basin/Pond and Retaining Wall

7.3.1 When the stability of gate chamber, stilling basin/pond, and retaining wall fails to satisfy relevant requirements, prestressed tendons may be used for reinforcement.

7.3.2 For gate chamber or stilling basin/pond reinforced with anchorage, the stability shall meet the requirements of SL 265 and SL 253.

7.3.3 For retaining wall reinforced with anchorage, the stability shall meet the requirements of SL 379 and the following:

1 The load borne by the retaining wall shall be shared by the dead weight of the retaining wall and the force resisting sliding applied by the prestressed tendons.

2 The number of prestressed tendons and the capacity of a single tendon shall be determined according to the stability calculation of retaining wall.

3 The optimal anchorage angle shall be determined according to the purpose, structural type and possible failure mode of the retaining wall.

7.4 Anchorage of Concrete Lining of Hydraulic Tunnel by Ring Prestressed Tendons

7.4.1 For a hydraulic tunnel under high internal water pressure, ring prestressed tendons may be arranged along the exterior edge of the concrete lining to bear all or part of the internal pressure through studies.

7.4.2 The structural design of prestressed concrete tunnel lining anchored by ring prestressed tendons shall comply with SL 279.

7.4.3 The elastic theory shall be employed for structural stress analysis of the prestressed concrete lining anchored by ring tendons, regarding the circumferential stress applied by the anchor as a component of the loads. When necessary, the finite element method or model test shall be employed for verification.

7.4.4 The design stressing force of a ring prestressed tendon shall be calculated with Formula (7.4.4-1) through Formula (7.4.4-3), taking account the radial preloading stress to be provided by the ring tendon and the deflector's friction loss at the tensioning end during tensioning:

1 Control values of tensioning stress may be calculated with Formula (7.4.4-1):

$$\sigma_k = 0.75 f_{ptk} \tag{7.4.4-1}$$

Where:

σ_k = control values of tensile stress of the steel strands during tensioning, N/mm^2;

f_{ptk} = characteristic value of steel strand's tensile strength, N/mm^2.

2 During stressing, the retained tensile stress after deducting the deflector's friction loss may be calculated with Formula (7.4.4-2):

$$\sigma_1 = (1-a)\sigma_k \tag{7.4.4-2}$$

Where:

σ_1 = retained tensile stress after deducting the deflector's friction loss, N/mm^2;

a = coefficient of deflector's friction loss, no greater than 8% of the control value of steel strand's tensioning stress.

3 The effective tensioning stress along the steel strand length may be calculated with Formula (7.4.4-3):

$$\sigma = \sigma_1 e^{-(kx+\mu\theta)} \tag{7.4.4-3}$$

Where:

σ = effective tensile stress along the steel strand, N/mm^2;

k = friction coefficient (wobble coefficient) of a steel strand, 0.0015 for a bonded steel strand and 0.0007 for an unbounded one;

x = length of steel strand from the tensioning end to the computation cross-section, m;

μ = friction coefficient between the steel strand and the sheath, 0.2 for a bonded steel strand and 0.032 for an unbounded one;

θ = angle included between the tangent of tensioning end and the tangent of the curved duct at the calculated cross-section, rad.

7.4.5 Ring tendons should be evenly arranged on the outer edge of concrete lining with spacing of 400 mm to 500 mm.

7.4.6 The strength grade prestressed concrete lining anchored by prestressed tendons should not be lower than C30. Only when the lining concrete reaches the design strength may the ring tendons be tensioned.

7.4.7 Anchorage slots shall be staggered on both sides of a tunnel for convenience of construction.

7.4.8 Unbonded steel strands should be arranged on the exterior of outer rebar. For bonded steel strands, ring corrugated duct shall be pre-embedded at the position of tendons, only when the lining concrete reaches the design strength may the ring tendon be installed and tensioned.

7.4.9 After completion of tensioning, the hole and anchorage slot shall be backfilled with non-shrinkage cement grout/mortar whose compressive strength shall not be lower than 30 MPa.

8 Design and Test of Safety Monitoring

8.0.1 The safety monitoring design of prestressed anchorage shall comply with the following principles:

1 Safety monitoring shall be focused on monitoring the safety of project construction and operation, providing guidance for construction and verifying the design.

2 Safety monitoring items in project construction period should be considered in combination with those in operation period.

3 Monitoring instruments shall be arranged at typical positions.

4 The selected monitoring instruments should be installed at the beginning of construction.

5 The number of monitoring instruments at each monitoring section should not be less than 3.

8.0.2 The monitoring instruments for prestressed anchorage shall be:

1 Durable, applicable, and adaptable to project environment, and their precision shall meet the project requirements.

2 In coordination with monitoring purposes.

3 Easy to realize both automatic and manual operation.

8.0.3 The design of safety monitoring shall include proposing technical requirements on installation, protection, and service of instruments.

8.0.4 The working status of prestressed tendons and the anchoring effect on the anchored objects shall be incorporated into the design of safety monitoring. Results of safety monitoring shall be basis for project safety assessment.

8.0.5 Monitoring items of the prestressed tendons shall include the following:

1 Actual elongation during tensioning and draw-in range during locking.

2 Anchorage force variation measured by force transducers installed on typical tendons.

3 Other items required.

8.0.6 Based on the anchored object and design requirements, one or more of the following items may be monitored:

1 Displacement or deformation.

2 Stress or strain.

3 Cracks.

4 Sonic wave velocity.

5 Ground water table and water quality.

8.0.7 Based on the type and grade of structures, the number of monitoring instruments shall be determined as per Table 8.0.7 and the following requirements:

Table 8.0.7 Percentage of tendons under safety monitoring for different structures

Type of anchorage	Slope			Underground structure			Hydraulic structure		
Structure grade	1	2	3	1	2	3	Dam foundation, Dam body	Gate pier	Gate chamber, Retaining wall
Percentage of monitored tendons /%	≥10	≥8	≥5	≥10	≥8	≥5	≥15	≥10	≥5

Note: The percentage of tendons under safety monitoring $=\frac{\text{the mumber of monitored tendons}}{\text{the total number of tendons}}\times 100$.

1 Monitored tendons in each anchorage zone or each monitoring section should not be fewer than 3.

2 Monitoring instruments for tendon element and anchored object should be arranged in the same monitoring section.

8.0.8 For special prestressed anchorage, specific monitoring design shall be conducted as required.

8.0.9 Arrangement and construction of monitored tendons shall be given priority.

8.0.10 When the measured stressing force exceeds 20% or is less than 10% of the design one, the cause shall be analyzed in time.

8.0.11 When the number of prestressed tendons is large and geological conditions of anchored object are adverse, the in-situ bearing test shall be performed before construction so as to determine the configuration and length of the fixed length. The requirements and methods of in-situ bearing test are shown in Annex A.

Annex A Test of Tendon Bearing Capacity

A. 0. 1 Either non-destructive test or destructive test shall be chosen for tendon bearing capacity test that shall be conducted in a geologically typical location.

A. 0. 2 The non-destructive test of the fixed length shall be performed as per the following principles:

1 Test tendons shall be representative.

2 It may not be conducted until the strength of grouting materials in the fixed length meets the design requirements. The application of stressing force shall be completed in 5 or more cycles and shall be held for 5 min at each stage. Observation shall be conducted. When the stressing force reaches 120% of the design one, the anchorage force of the fixed length may be deemed to meet the design requirements.

A. 0. 3 The destructive test of the fixed length shall be performed as per the following principles:

1 It shall not be conducted for the mounted stressing head.

2 The tendons used for destructive test may be shorter than the mounted stressing head, however, the length of the fixed length shall meet the design requirements.

3 The application of stressing force shall be completed in cycles as specified by the design of stressing procedure. Each stage's stressing force shall be held for 5 min, and corresponding observation shall be carried out. When continuous displacement occurs in fixed length or 30% of the steel wires/strands are fractured, the prestressed tendon may be deemed to be in destructive state.

4 If Subclause 3 above is not applicable when the stress of anchorage reaches the ultimate tensile strength of steel, the anchorage force of the fixed length may be determined based on the relation-curve of the measured load and deformation.

5 The safety degree of fixed length shall be verified by test results. And the configuration and length of fixed length shall be adjusted accordingly.

6 Protective measures shall be taken to protect personnel and property from injury and damage when the destructive test is under way.

Explanation of Wording

<table>
<tr><th>Words in this Standard</th><th>Equivalent expressions in specific situations</th><th>Strictness of requirement</th></tr>
<tr><td>Shall</td><td>It is necessary</td><td rowspan="2">Required</td></tr>
<tr><td>Shall not</td><td>It is not allowed/permitted
It is unacceptable</td></tr>
<tr><td>Should</td><td>It is recommended
It is advisable</td><td rowspan="2">Recommended</td></tr>
<tr><td>Should not</td><td>It is not recommended
It is not advisable</td></tr>
<tr><td>May</td><td>It is suitable
It is desirable
It is preferable</td><td rowspan="2">Permitted</td></tr>
<tr><td>Need not</td><td>It is unnecessary
It is not required</td></tr>
</table>